Disclaimer
Despite all our efforts and those of HarperCollins in Australia, owner of Angus & Robertson, who we thank for their help, we could not find information on Alexander Carr Bennett's heirs. They are free to contact us about this new edition.

Talma Studios
60, rue Alexandre-Dumas
75011 Paris - France
www.talmastudios.com
contact@talmastudios.com

Cover image: © Oliver Sved I Dreamstime.com

ISBN: 979-10-96132-00-3
EAN: 9791096132003
© All rights reserved

ELECTROCULTURE

THE APPLICATION OF ELECTRICITY TO SEEDS
IN VEGETABLE GROWING

BY
A. CARR BENNETT

ST. QUENTIN EXPERIMENTAL NURSERY
WENTWORTHVILLE, N.S.W.

With 32 Illustrations

First Edition:
ANGUS & ROBERTSON LTD.
89 CASTLEREAGH STREET, SYDNEY
1921

CONTENTS

Chapter	Page
FOREWORD	8
INTRODUCTION	9
I. THE DISCOVERY	15
II. THE BENNETT METHOD	23
Electrification Table	28
Renewal of the Current	31
Electrification from the Lighting System	34
Other Methods of Seed Electrification	35
III. AFTER-TREATMENT	37
Laying out the Garden	37
Watering	39
Mulching	44
Manure	44
Rotation	49
Seedbeds	49
Testing for Germination	51
Transplanting	52
The Root System	54
IV. SPECIAL VEGETABLES	57
Beetroot	59
Cauliflowers	60
Celery	61
Cucumbers	62
Onions	66
Parsnips	69
Rhubarb	72
Strawberries	72

Tomatoes	73
White Turnips	76
Table of Monthly Sowings	78
V. MISCELLANEOUS HINTS	**81**
Economizing Ground	81
The Value of Trellises	82
Piping and Hose	84
Application of Current to Growing Crops	87
Weeds, Pests and Remedies	89
Growing without Gardens	90
Specialization	92
A Larger Sphere	93
VI. THE TESTS AT ST. QUENTIN NURSERY	**97**
APPENDICES	**109**
I. Evidence from Duntroon	109
II. Recipes	113

ILLUSTRATIONS

	PAGE
Visitors Inspecting Plots at St. Quentin Nursery	13
Drumhead Lettuce grown at Armidale	20
Epicure Beans grown at Lakemba	24
Medical Coil used by the Author	30
Seed-drying after Electrification	32
Electrified American Wonder Peas	36
Sprouting a Potato Set	41
Test Beds at St. Quentin Nursery	46
Seedlings planted round an Intense Irrigator	50
Cabbage plant twelve weeks old	53
Rootlet System of Electrified Tomato	55
Blood-red Beets	56
Pumpkin, Cucumbers and Tomatoes	63
Earliana Tomatoes	65
Parsnips, Carrots and White turnips	67
Electrified and Non-Electrified Cucumbers	68
Cucumbers on the Vines	70
Rhubarb, Giant Red	71
Tomatoes, Duke of York	75
Tomatoes. Burwood Prize	77
Silver Beet	80
White Nepaul turnips	83

Canadian Wonder French Beans	86
Treating a Sick Cabbage	88
Iceberg Lettuce	95
Growing Cucumbers without yard space	96
Seeding Lettuce	98
Tomato Seedlings	103
The Electrified Melon Patch	104
Giant bunch of Radishes	107
Tomatoes ready for Market	108
Young Gums at Duntroon	111

FOREWORD

After WW2, agrochemicals replaced all other techniques which have been tested and used for decades or even longer in some cases. One of the most promising was electroculture, which consists of applying electricity to seeds, in order to stimulate the growth of plants.

There are different ways of doing it, and over nine years Alexander Carr Bennett developed a method which proved highly successful even in conditions of poor soil or little irrigation.

Almost a century later, we are pleased to once again be able present this information to everybody, including farmers, gardeners, agronomists and anyone interested in organic food, health and environment.

Patrick Pasin
Publisher

INTRODUCTION

At no time in the world's history has there been greater need of improvement in gardening methods, especially in those which induce a speedier and intenser growth of vegetables. The decreased output of foodstuffs, resulting everywhere in increased living-costs, lays upon every householder the duty of producing at least a part of what his family eats. The part easiest to produce—at any rate in and near the great centres of population, where land is to be had only in small allotments—is the vegetable food so necessary for continuous good health. Of this Australians eat far too little, even in times of plenty.

In Australia the conditions are ideal for the growth of almost every known vegetable. As a rule the father of a family cannot make clothes or footwear for his family, or grow animal food for it; but if he has a patch of land he can grow its vegetable food: and, in an emergency, vegetables are capable of replacing animal food altogether, while at any time they provide the body with a much larger proportion of the substances necessary for complete health than is generally understood. Land, of course, is essential: and nowadays land is costly. One way of getting over this difficulty is to devise means of making it cheaper—but that we leave to the politicians. Another way is to make it possible to get more vegetables and better vegetables, and to get them more quickly, out of whatever area the average man may have at his disposal. It is claimed for the system described in this book that by its aid no back-yard is too small, no soil too poor, to grow vegetables in such quantity and of such quality as will materially lessen a family's food-bill: while any man lucky enough to have half an acre at his disposal can grow not only all the vegetables needed for a large family, but can earn a full living wage from the sale of the remainder.

This is a big thing to claim. But it is being done at the present moment within a few miles of Sydney, and anyone with enthusiasm enough to visit the experimental farm at Wentworthville can verify this for himself. Moreover, the father of the family need not give up all his spare hours to gardening. Once he has broken up the soil, laid out the beds and seen to it that water can be administered to them handily, he may leave the rest to the mother of the family, who will find the burden of her household duties augmented very little by the addition. That is the special advantage of a system which aims, not at cultivating large areas or employing elaborate stimulating agencies, but at getting the most by the simplest means, and at the highest speed, out of every square inch of ground. The preliminaries (which, of course, are the essence of the system) once gone through, the chief labour is watering the plants: and the watering of even half an acre with a hose is no heavy task, while the watering of the plots and boxes (which though tiny, will easily grow enough to supply the average family) is a job the busiest housewife will take pleasure in.

The Bennett system consists in the stimulating of fertile vegetable seed, *before* sowing, by passing through it weak currents of electricity, and in the use of plentiful waterings to hasten its growth *after* it has been sown. In this book it is proposed (a) to give the history of the author's discovery of the principle, and narrate his first failures and gradual approach to success, (b) to describe in the simplest language his method of electrification, the subsequent treatment, and the dangers to be guarded against, (c) to give the result of continuous and careful tests carried out at his experimental farm at Wentworthville and elsewhere, showing what can be done on poor soil with ordinary care. Moreover, as the system has for some time been made public through the columns of the Sydney *Mail*, the difficulties actually experienced by the readers of that journal who are giving it a trial have been brought under the author's notice, and have been taken account of in the course of writing the book. The idea of stimulating the growth of plants by electricity is not a new one. Between 1885 and 1905 a Finnish professor experimented with strong currents conducted through wires above growing plants; the latest methods in electrification of this type call for currents of high voltage. Other methods, still in the experimental stage, either collect and distribute atmospheric electricity, or pass currents through the soil between buried electrodes, or (in greenhouses) expose the plants to light from an electric arc. More recently, attempts have also been made to electrify seed in bulk (usually cereal grain), by soaking the grain in various chemical solutions and then placing it in the path of a current between electrodes. These methods are mentioned lest it should thought that the author is deliberately ignoring them. But it must be understood that the type of Electroculture described and advocated in the following chapters is an original discovery, made without knowledge of earlier experiments with seed. It differs from all the abovenamed methods, except the last-mentioned, in that it deals with seed; and from the last-mentioned in that it deals with large or small quantities of seeds of all kinds by means of a low-priced apparatus that anyone can handle—and that the use of chemical solutions and other complications is rendered unnecessary.

The Bennett system of Electroculture is practical beyond all others. The author discovered his system for himself: but from the moment he realized the value of the application of electricity to growing things he studied and experimented with all the other system, and found that they are not commercially workable under Australian conditions. They are too costly for any but the rich and the grower on a large scale. The Bennett system, on the other hand, starts at the beginning by energizing the life germ in the seed, and by using small and simple forms of apparatus adds only slightly (one might almost say infinitesimally) to the cost of cultivation. It is, therefore, adapted to the needs of the small farmer and the suburban grower, and to the family man whose sole object is to provide his family with more and cheaper food.

Visitors Inspecting the Plots at St. Quentin Nursery, Wentworthville, N.S.W.

CHAPTER I

THE DISCOVERY

WALKING one day down the main street of Lismore in northern New South Wales, the author's eye was caught by a map exhibited in an office window, round which others were already crowding. It showed the fourth official subdivision of the Dorrigo lands, at that time widely advertised by the State Government as an excellent "closer settlement" area. He became a convert at once—the lure of the land gripped him; a block was applied for, and in due course allotted; and before long he became one of the crowded population of the little Dorrigo township, on the summit of one of the finest wooded plateaux in Australia.*

The extraordinary fertility of the Dorrigo, which grows in some places impenetrable thickets, in others huge forest giants closely packed together, and everywhere a wealth of festooning creepers and ferns of all kinds, was from the first both a surprise and a puzzle to the newcomer. Chemical fertility of soil accounted for a good deal, as did the excellent

---

* A line drawn on the map from Armidale in New England to Coff's Harbour on the coast crosses the plateau.

rainfall. But there was more in it than could be thus explained. A study of local statistics soon showed that the years of great rainfall were not the years of great growth. Then memory recalled a hint given in the first days after arrival : "These terrible thunderstorms" a neighbour had said, "are one of the worst features of the Dorrigo." Perhaps—perhaps not. It is a common-place that a thunderstorm "clears the air"—which simply means that, when once its violence has been spent, its after-effects are exhilarating to every living thing. Might not this exhilaration, recurring after every storm, and not necessarily proportioned to the violence of the storm, have something to do with the exceptional verdure of the Dorrigo?

This train of thought once started, it next became necessary to prove or disprove the assumption that Electricity was the power that accomplished these wonders. It was easily shown that storms of an electric character were more frequent in the years of exceptional growth. Observation also soon demonstrated that terrific storms, whilst doing damage, did not increase the growth to any extent, whereas a mild disturbance with perhaps little rain did the maximum of good. The author gradually evolved a theory to account for this, which may be here stated.

In a severe storm the concentrated violence of the lightning-stroke will blast a huge tree: yet the self same current will benefit the foliage surrounding it. Why? Possibly the atmosphere acts as a regulator, reduces the amperage or amount of the current, and distributes it over the surrounding growths in a mild form which acts as a tonic reviver. Only at the spot, which is struck by the lightning is there a great amount of current.

When the storm is severe, and what we speak of as a "bolt" descends, it rends, blasts, and destroys the object it is centered on. It is, in fact, intense electric energy concentrated in a very small area. Beneficial storms are those where the lightning, though severe, does not make a particular point its objective. These storms, and mild ones, scatter the electricity throughout the atmosphere—which, at such times, is always of a slightly damp character—and *the atmosphere, acting as a conductor and regulator combined*, delivers the electricity in a mild current through the leaf-cells and the root-system of the plants.

In certain districts the formation of the country, or its metalliferous nature, attracts the electricity in the atmosphere; consequently these areas become noted for productivity which often enough is attributed wrongly to the soil or rainfall.

Thus from thought to thought were the footsteps of the enquirer placed firmly on the right track, and he started out to copy Nature. For some time the results were not at all satisfactory: he did not understand the delicacy of the seed-embryo that he sought to treat. It required more lessons from Nature, and more scrub[1] walks, before a proper system was evolved. Experiments were carried out with primitive apparatus, the trials being kept secret until definite results had been obtained.

---

1. "Scrub" is the technical name of the great soft-wood forests of eastern Australia. The word "forest" is technically confined to groups of hard-wood trees. Thus "scrub" as used here has no connection with undergrowth: it includes, indeed, cedars and pines and other big timber.

The "medical coil" was ultimately selected as an excellent apparatus for the administration of mild electrical currents. Years of experimenting have proved that it is the rate of current-flow that we have to take into consideration with seed electrification. Different voltages give similar beneficial effects, so long as the amount of current is low. This makes the medical coil essentially the proper apparatus for seed electrification; for the flow of current through this medium is so small that it has to be measured by milli-amperes, or thousandths of an ampere.

For a considerable time, however, ill-luck dogged the experimenter. The tests were carried out on one class of seed—lettuce—for it was recognized that, success once obtained with it, a like success with others would soon follow; further, by concentrating on one variety the field of enquiry was narrowed. Lettuce-seed of average germinating quality was selected from packets obtained from reputable seedsmen. In the earlier tests it was in every instance planted with either check-rows or beds of untreated seed for purpose of comparison. The primary current was first tried for fifteen minutes at scale strength of 80; no benefit worth mentioning could be noted. Realizing this, the experimenter turned to the use of the secondary current. The tests gave practically the same result. The combination of primary and secondary currents was then tried at the same strength, and the results were decidedly better. Treated seed germinated several hours before the untreated; the plants maintained the advantage they had gained, and matured considerably ahead of the check plants. Not much publicity was given to the experiment at this stage; one swallow does not make a summer, or one success prove a theory correct. The lettuces were, however, inspected by several neighbours.

Later, the vicissitudes of Fortune took the experimenter to Armidale. Securing a suitable experimental plot, he decided to delve further into the subject. He was sure that he was still not supplying the current in the natural manner or proportion. Though the speed of germination had been increased, and the number of plants that came to maturity augmented, he wanted still better results and higher germination. The question of the class of current had now been definitely settled to his satisfaction, both by experiments, and by reading of the discovery made in India that the electricity brought down by rain was sometimes of a positive, sometimes of a negative character. The research, therefore, narrowed itself down to discovering the correct amount of current, of a pulsating type, to give the very best results. This could only be arrived at by exhaustive experimenting, using lower and higher currents, noting the different effects and trying them over different periods of time. It soon became apparent that "fifteen minutes at 80" was on the high side for lettuce seed. After several further tests "ten minutes at 50" was selected as giving the best results with lettuce; and to this day the experimenter cannot better it.

Parsnip seed was next dealt with. From the start the experimenter was more successful, and arrived at "fifteen minutes at 45" with half the trouble the lettuce had given. Beet, however, required a current at 80 and a time limit of fifteen minutes. Similar experiments were made to ascertain the correct

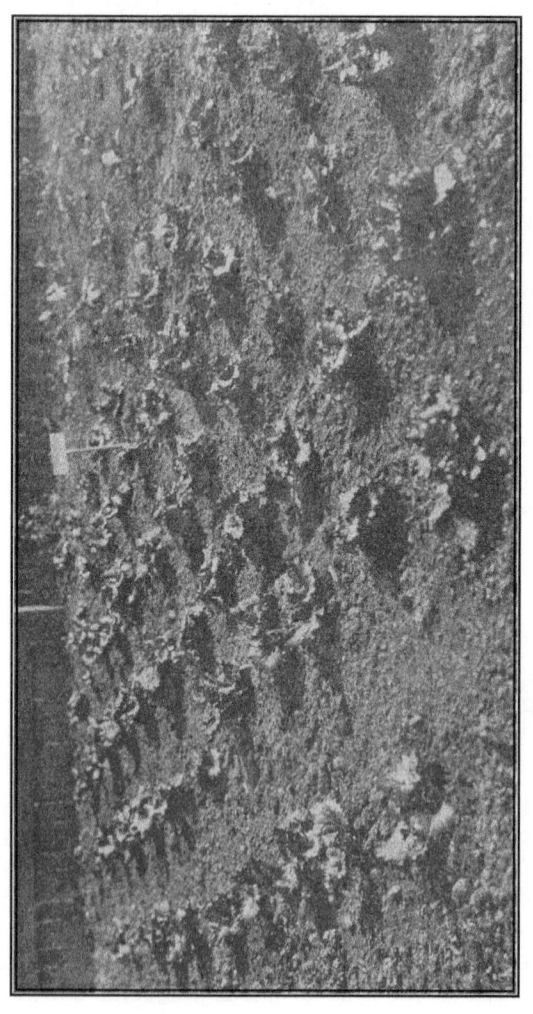

Drumhead Lettuce Grown at Armidale, N.S.W. from Electrified Seed. Not one missed, and all hearted within six weeks of planting the seed.

time necessary for thorough saturation of each kind of seed—peas, for instance, required a much longer soaking than lettuce. Tests were now carried on with many other vegetables, and gradually evolved the scale given on page 28, which, though not infallible, is a good working guide to seed-electrification by means of the medical coil. The scale gives only a relative strength, as the amperage given by this class of apparatus is so small that it cannot be measured by that standard. Many difficulties had to be surmounted, and numberless failures were written off. The main difficulty was that, in common with most enquirers, the experimenter looked past the self-evident and tried more complicated and more complex devices; further, he failed to appreciate the great difference between Electroculture and Electrocution. Even as recently as during the first writing of this book he killed a quantity of seed by applying too strong a current—he was in a hurry, and had placed in the coil a new cell of a different make. Familiarity with a subject often includes carelessness, and it is so in this case. He now makes a point of applying the current to himself, by means of hand electrodes, before applying it to the seed. The sensitive nerves of a human being soon note anything wrong with the current, and easily discern if that current is stronger than it should be at a given point.

Another cause of failure in the earlier tests was the heaping together of the seed and uneven distribution over the surface of the plate. Uneven distribution meant uneven electrification, and uneven electrification gave uneven results.

After about thee years in the Armidale district, the experimenter removed Sydney-wards, and the tests were continued at Lakemba on the Bankstown line. Here again the use of the current and the growth of the vegetables attracted attention. The discovery now received more general recognition, and articles dealing with it appeared in Sydney newspapers at frequent intervals. The principal success at Lakemba was with cucumbers, epicure beans and tomatoes. Enquiries began to pour in from all quarters, and many applicants for electrified seed had to be refused.

At this stage Mrs. Mary Bennett purchased an acre of land at Boronia Park, Wentworthville, and offered it to the experimenter to be run as an Electric Nursery — the first of its kind in Australia. The area was covered with dense couch-grass, and had to be ploughed and stumped. The next operation was hand-hoeing out the couch, which was stacked and burnt; the land was then thoroughly drained and laid out in small beds, and named "St. Quentin Electric Experimental Nursery." In chapter VI are described the results obtained in this area of poor land, without the aid (roughly speaking) of artificial or stable manure, and relying solely on the stimulation received by the seed from the electrification process.

CHAPTER II

THE BENNETT METHOD

THE apparatus required is simple and inexpensive, being an ordinary three-inch medical induction coil such as can be purchased at any electrical supply store. This, with patience and common-sense, is the only thing absolutely needful at the start. If the enquirier has read such a book as *First Studies in Plant Life in Australia*, by William Gillies, he will find it helpful.

Before setting out to electrify seed we must realize that the germ we are dealing with is a very delicate one, and that if we apply too much current we shall kill the little germ: whereas if we apply the correct quantity its vitality will be increased.

The first thing is to make acquaintance with the medical coil and discover the exact use of its several parts. These comprise (a) the induction coil with its contact-breaker, (b) two flexible insulated wires with terminals, (c) a metal foot-plate,[2] (d) two

---

2. If the foot-plate supplied with the coil is too small, or has been lost, another may be substituted. The substitute should be of zinc, copper, or any metal that is a good conductor, and must be insulated by placing it on glass or porcelain (at need, a dinner-plate will serve). It should be firmly attached to the flexible insulated wire.

Epicure Beans. Height 5 feet. Grown at Lakemba from Electrified Seed. Fourteen pounds of beans were picked from a few vines at one pulling.

handles (called "electrodes") with holes into which the terminals of the wires fit, and (e) fixed on the wooden framework that carries the coil, three plug sockets and a switch (Some coils have screwed terminals instead of the sockets, in which case terminals attached to the wires are adapted to fit into the screwed terminals). At the right-hand side of the coil will be found a pull-out current regulator, and below it a scale that indicates the intensity of the extra current induced by drawing out the regulator. In the box beneath is a No. 6 Columbia or Mesco dry cell, which is attached by wires. (When the current has been exhausted, purchase another dry cell and attach it yourself).

To start the current, in the first place two terminals, one from each flexible wire, should be placed in two of the sockets. Using sockets Nos. 1 and 2. we get the primary current; with Nos. 2 and 3, the secondary current; with Nos. 1 and 3, both currents. The last is the arrangement found most useful for our requirements. Place, therefore, a terminal of one wire in socket No. 1, and a terminal of the other in socket No. 3; connect the other terminals, one with the foot-plate and the other with the wooden-handled electrode; and throw over the starting switch from left to right. If the armature does not begin to hum immediately, turn the small screw at the left of the coil until the humming begins. The current is now on.

It being next to impossible to pass a current through a heap of dry seed, it will be necessary to soak the seeds before electrifying. The soaking must be thorough (see Table on p. 28). When the seed is sufficiently moist, place a damp cloth on the foot-plate, connect up the battery as already described, lay the damp seed in a thin even layer on the moist cloth, and place another moist cloth over it.

The starting switch is now thrown over, and the actual process of electrification commences. Pass the electrode with the wooden handle backwards and forwards above the seed, with a stroking movement, the metal portion just touching the cloth. The current should be of strength 10 on the scale to begin with, and be increased by pulling the regulator out gradually until you have reached the maximum strength shown on the table for the particular variety of seed being treated. During the last third of the operating time the current should be at full strength.

See that the current is evenly distributed, as otherwise the treatment will result in an uneven growth in the plants. Don't pull the regulator out suddenly, as the seed would then come under a heavy current from the very beginning. Remember that the desired effect is obtained by gentle, gradual application. Heavy currents kill.

As the sowing of damp seed (especially small seed, such as lettuce) is extremely difficult, it will be found advisable to dry the seed after electrification. There are many methods of doing this. The writer has made a dryer out of a butter-box, heated by a small lamp underneath. Of course the seed can be easily dried in the sun on a fine day. If this is dried by artificial means, it must not be subjected to undue heat, otherwise the germ will in all probability be destroyed. Though the effect of the electrification does not pass off for some time (in some experiments seed has been sown twelve months after electrification) the best time to sow electrified seed is as soon as possible after treatment.

It will be found more satisfactory if you electrify only the quantity of seed you are ready to plant. The ordinary methods of sowing can be followed. See that the seed is fresh to begin with, for, though the current will hasten germination, it cannot bring the dead to life.

Various solutions, including common salt and nitrate of ammonia (about a teaspoonful to a pint of water), have been tried as soaking agents. The saline solution is a good conductor of the current, but has a detrimental effect on the growth. Pure water gives the best results. Still, if beet seed is soaked in a weak solution of liquid manure before being treated, it will help it materially.

ELECTRIFICATION TABLE

As many enquirers ask for time-tables for the soaking and treatment of seeds, the following guide is given. It is not claimed that it is infallible; but judged by the author's experience it is mighty near it.

Time-table for Soaking and Electrifying Seed

Seed.	Time to Soak.	Current and Elect.-time.
Beans	12 hours	15 min. at 80
Beet	3 hours	15 min. at 80
Cabbage	30 min.	10 min. at 75
Carrot	15 min.	10 min. at 80
Cauliflower	30 min.	10 min. at 65
Celery	10 min.	10 min. at 60
Cucumber	12 hours	15 min. at 80
Kohl Rabi	30 min.	10 min. at 75
Lettuce	15 min.	10 min. at 50
Melons, Pumpkins and Squashes	12 hours	15 min. at 90
Parsnip	15 min.	15 min. at 45
Peas	6 hours	5 min. at 80
Potatoes[3] (apply direct to cut portion with foot-plate)		20 min. at 80

---

3. These do not need to be soaked.

Radish	30 min.	10 min. at 75
Rhubarb[4] (apply direct to root for)		15 min. at 80
Tomato	15 min.	15 min. at 50
Turnip	30 min.	10 min. at 75

Farm Crops

Maize: Soak in pure water till thoroughly moist—no salts. Electrify 20 min. at 80.

Grass Seeds: Cocksfoot, Rye, Paspalum, Sudan and other grasses. Soak 20 min. Electrify 10 min. at 50.

Peanuts: Shell and soak for 6 hours. Electrify 5 min. at 80.

Pigeon Peas: Soak for 6 hours. Electrify 10 min. at 40.

Wheat: Soak as usual in bluestone; apply current by means of copper plates on either side of bag containing seed, after insulating cask by placing on piano insulators. Time suggested is 10 min. at 70, but the author's experiments with wheat are still in progress.

Note—In every case the figures deal with electrification by a medical coil, and the maximum strength is given. Start with a low strength, and gradually increase till maximum is reached.

---

4. These do not need to be soaked.

THE MEDICAL COIL USED BY THE AUTHOR.
THE SCALE IS PLAINLY VISIBLE. ON TOP IN LID IS FOOT-PLATE USED FOR SMALL QUANTITIES OF SEED AND HAND ELECTRODE USED IN STROKING. THE CORDS ARE IN HOLES 1 AND 3; THE REGULATOR CAN BE SEEN AT SIDE TO RIGHT OF THE COIL.

Now the above type of medical coil usually receives its energy from a dry cell. The cell recommended is a Mesco (A. Size) or a Columbia No. 6. It is recognized that these dry cells give a maximum current when first switched on; the strength then declines for a few hours, after which the current will flow for a considerable time with but a small decline. (It is, of course, possible to place other makes of dry cell in the medical coil, but the behaviour of the current might be different.) Seeing, then, that it takes these cells a certain time to reach a fairly uniform strength, it is advisable, if the battery is used for seed electrification whilst still fresh, to vary the table slightly, giving in each instance five points less. The scales given in the foregoing Table has been computed for the "uniform" strength of the battery, after the cell has been at work for, say, two hours. *Most failures result from the use of too strong a current*, and it is necessary to exercise great care in administering a current of the right strength, or the minute germ will be injured or destroyed.

Renewal of the Current

As the user will need to place another dry cell in position, after the one obtained with the battery is exhausted, a note of warning is here sounded against using any battery but a Mesco A. or a Columbia No. 6, or a Red Seal of the same type. The Table on p. 28 has been carefully worked out for these particular cells; other cells give either a greater or lesser current, and thus render the Table an unsafe guide. If a wet cell is used, the Table will only be a comparative guide, inasmuch as the

SEED-DRYING AFTER ELECTRIFICATION, WITH THE HELP OF A BUTTER-BOX, A SMALL BEDROOM LAMP AND A PIECE OF WIRE SCREEN.

figures given will not be applicable. This should be no deterrent, however, to any person possessed of electrical knowledge; for they will realize that, as long as the current is low, there is little danger of destroying the seed's power of germination by harming the delicate organism. In placing a New Columbia or Mesco cell in position, be careful to attach it in precisely the same way as the one that was in the battery when you bought it, and see that all screws are tight and nuts well screwed into position. An improperly attached cell will not give the desired results.

Many of the medical coils sold contain no instructions for their use; but, if the advice in this book is carefully followed, there will be no failure in seed electrification. It is possible to revive the dry cell when run down by soaking it in a solution of sal ammoniac, which it will absorb. Before soaking, a few small holes should be punched through the case of the cell; these, after soaking, must be completely sealed by means of solder. It is advisable to do this only when the user cannot readily obtain a new cell. An experienced person can test the strength of a dry cell by applying his tongue to the two nuts on top. Dry batteries should not be left out in the moist air, or they will deteriorate. Ask your dealer to try the cell on his test bell, or in some other way, before you buy; otherwise you may purchase a run-down and useless one.

The method of electrification here recommended for the treatment of vegetable seeds—placing them on a metal tray with a damp cloth interposed—is considered the best, as it has given the author unequalled results. A friend, expert in electrical science, was of opinion that the seed could be treated just as well whilst in the soaking solution, if the vessel containing it was a non-conductor. As this would save the plate treatment, it was decided to try it. The results were not satisfactory; the experimenter set out to find the reason, and was not long in arriving at the conclusion that the current was so much altered by passing through the fluid (whether nitrate of ammonia solution or plain water) that the table could not apply. It was decided to adhere, in the case of seeds already thoroughly tested and scaled, to the first method, which has given such splendid results. If treated while in the solution the seed should, in the opinion of the writer, receive a third less time and a third less strength than is recommended in the table.

ELECTRIFICATION FROM THE LIGHTING SYSTEM

Those who use electricity for lighting purposes may wish to take the current for seed-treatment from that source. This is quite practicable; but, as the current in lighting mains is from 100 to 240 volts, some means of transforming it to a lower power is necessary. The Rheostat, a piece of mechanism which can be purchased at your electrical dealer's, will solve the problem. This instrument is designed to vary electrical resistance between points. In no case is it possible to treat seeds direct from the heavy lighting systems; that would mean electrocution of the tiny germ. The current, no matter what its source, must be mild when applied to seeds. Some idea of the difference between the strength of a lighting current and that of the current we use for the electrification of seeds may be gained by comparing the following particulars. An ordinary incandescent lamp of 16-candle-power, supplied at a pressure of 200 volts, takes a current of a quarter of an ampere. An arc lamp, similar to that used for street lights, takes a current of 8 amperes. A ten-horse-power motor, working at full load from a 200-volt main, takes 50 amperes. But in the electro-medical apparatus, such as the coil we recommend, the current has to be measured by thousandths of an ampere.

OTHER METHODS OF SEED ELECTRIFICATION

Many other methods of treating seed with an electric current present themselves to an experimenter. Among others, a current may be passed through a quantity of seed in a porcelain vessel filled with water. If this method is adopted, it is necessary to remember that the table given above will not serve as a guide, for the presence of water greatly alters the current. The current must be evenly diffused through the seed, and in no case must it be applied too strongly. The time-limit may in certain cases be extended, but the current must then be milder.

Large seeds, such as beans and peas, are perhaps better treated whilst soaking. This is made easy by the introduction of a zinc plate at either side of the vessel, which must have previously been placed on some non-conductive material to prevent the current from escaping.

When electrifying potato sets, the best method is to place the cut portion on a plate to which the current is applied. After treatment, rub the cut in wood ashes before sowing.

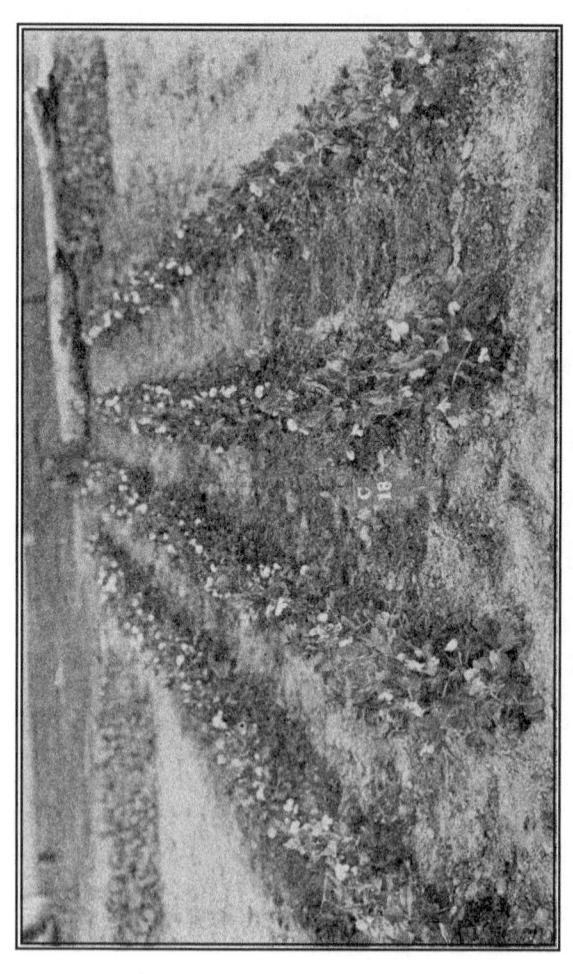

ELECTRIFIED AMERICAN WONDER PEAS.
THIS BED GAVE MARVELLOUS RETURNS—THREE HEAVY PICKINGS, AND THE PODS ALL WELL FILL, MANY WITH EIGHT PEAS TO A POD. NO FERTILIZER USED.

CHAPTER III

AFTER-TREATMENT

LAYING OUT THE GARDEN: GENERAL HINTS

The laying out of the garden is a most important affair; if it is properly designed, the labour required will be fifty per cent less. An oblong piece of land lends itself best to our requirements, and a path of at least three feet wide should be made right up the centre to leave room for wheeling a barrow comfortably. The water-main should be taken down the middle line of the garden to its centre, and stand-pipes and taps placed at such distances that every part of the area can easily be watered without any carrying. If the main path is bricked, asphalted or gravelled, it will always keep dry in t he wettest weather. The beds should be oblong. Small beds will prove the handiest, and every bed should be readily accessible. Large beds do not suit the electroculturist.

If it is possible to choose the site of the garden, an easterly slope is most to be desired. If the land is bumpy, an effort should be made to grade it as nearly level as possible; this levelling up will prove a boon in the future, as channel irrigation can be practised on a small scale. Water of some kind must be easily available if the electroculturist is to succeed. A sprinkler system will prove a great blessing, and in one season will repay the outlay. If you use a hose, it should be long enough to reach all over the garden from the standpipes.

The plot should be, as far as possible, clear of trees; they suck up moisture, and deprive other vegetable growths of nourishment. If the position is windy, a breakwind trellis covered with passion-fruit vines will be the best means of coping with the difficulty. These vines keep green, provide shelter, and produce a valuable and easily marketed crop. In climates unsuitable for passion-fruit some other creeper of dense growth may be substituted. The soil should be worked up as fine as possible, and stable manure, if obtainable, should be incorporated before sowing. Well-rotted stable manure is the best of all fertilizers, and improves the quality of the soil. Failing this, suitable fertilizers, such as blood and bone manure, will prove an excellent substitute. The land should be well drained; but the writer has not found trenching advisable in most of the soils round Sydney. A couple of small pits, one for stable manure and the other for green manure, will be most serviceable. Tops and green growths should not be wasted but be placed in the pit to decompose, and later be dug into the beds. If "Intense Irrigators"[5]

5. Intense Irrigators are perforated earthenware pots which can be sunk at convenient intervals in the bed, and form a convenient method of administering water or artificial liquid manure direct to the roots. They can be filled from a drain or a hose. They are especially useful for small beds and boxes, as the roots receive all the moisture or liquid manure supplied.

are placed in the beds of lettuce, nitrate of soda (a valuable stimulator) can be readily introduced to the roots; this will help to push the hearts along.

Cucumbers and melons should be placed in such a position that they will not overrun other beds, as they quickly become a nuisance.

In planting seed the row principle will in most cases be found best. The rows can be placed close together, and worked with a small rake or Norcross cultivator. Broadcast beds are more difficult to keep clean, and do not admit the sun so readily to the plants. Some sort of a shed in which early plants may be started will prove a great convenience and a money-maker. It should be the aim of the electroculturist always to have another bed coming into work when one is going out, and thus to keep up a constant supply of all vegetables in season. It will often be found practicable to start the new crop while the other is maturing. Take, for instance, a bed of lettuce that is nearing pulling time; in the season cucumbers can be started between the lettuce, and will be running by the time the lettuce have cut out. By adopting this system there will be no empty spaces, and all the ground will be working.

Watering

It is well to bear in mind that when we electrify a seed germ we hustle it into great activity, and therefore must supply sufficient nutriment and water to allow of rapid growth. A starved plant can not possibly retain its health and make vigorous growth, no matter what start it may have received at the beginning of its life.

As the main constituent of every vegetable is water, it is absolutely necessary that water should be supplied—and more so to the electrified product than to the ordinary unenergized one, since the former requires more water in less time.

The vital importance of a proper water-supply is often overlooked by persons who are starting gardening for a livelihood. Too much emphasis cannot be placed on the necessity of a good supply. Slightly damping the surface is not watering, and does more harm than good. It may be truthfully asserted that ninety per cent. of cultivated plants never reach proper development owing to an insufficient supply of water.

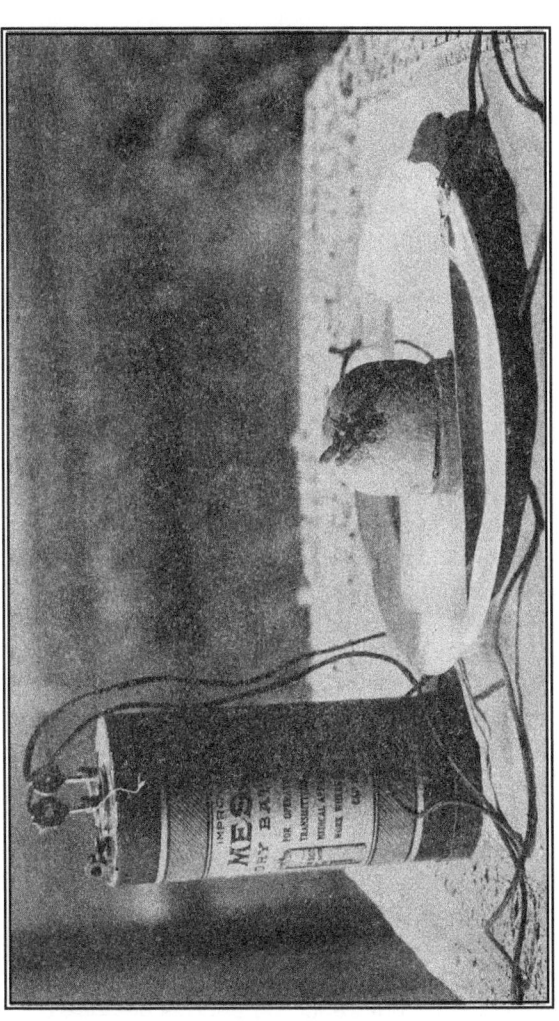

Sprouting a Potato set on a zinc plate by direct induction from a dry cell. The set sprouted 1½ inches in 24 hours, and threw out strong rootlets. When planted, it gave an increased crop.

The French Maraichers—the gardeners *par excellence*—realize this fact, and make ample provision for water supply. During the hot months of the year they water continually, not dribbling the water on the surface but thoroughly soaking it. Occasional showers of rain are of very little benefit. You may laugh to see a man watering a garden after or during a shower of rain; but that is often seen in a French garden—and these gardeners make up to a thousand, or in special cases even fifteen hundred pounds per annum per acre. The electroculturist can learn much by adopting the Maraichers' watering methods. On a hot day in midsummer as much as 26,000 gallons is often used on a French garden two acres in extent.

A small pipe service should be avoided. The author has laboured under the disadvantages of ¾-inch pipe services, and knows their shortcomings. Bigger pipes cost more in the beginning, but later on pay for themselves a hundred times over. *No pipes in the main line should be under two inches in diameter, and none under 1½-inch should be used anywhere in the garden.* The pipes should be so placed that every foot of ground can be readily supplied. A good plan would be to have revolving sprinklers of the "Rainmaker" type, fixed to permanent pipes at intervals that will allow of the whole garden being sprinkled. The turn-off taps should be placed down a central path, so that any portion or the whole of the garden may be watered together or separately.

Another good method is to have raised beds, and so allow the water to run between each couple of rows. If this method of trench watering is practised, the garden will need some sort of grading. The sprinkler method, however, will be found to give excellent results so long as the beds receive a thorough saturation. A good watering at intervals is better than a slight one every evening. If water is taken from a well or an underground source, you should first ascertain that it does not contain any harmful salts.

Both underneath and surface watering have their uses, but the writer believes that top-watering is absolutely essential, for the same reason that makes a human being take baths—to free the pores of accumulated and choking dirt.

Much has been written on the value of irrigation-meaning the application of water to growing crops by means of drains. Now the writer contends that in a very dry climate any irrigation that does not flood the leaf foliage is a failure. "Water applied to the roots is all right, but it must be supplemented by top-watering. In a district where there is a fair rainfall, drain irrigation is satisfactory.

After studying the habits of plants, it is easily conceivable that some such system as the Skinner[6] represents the ideal method of applying water. Top-watering relieves the leaves of dust and dirt that would otherwise clog the minute cells, and allows these cells to collect carbon dioxide-a most valuable plant food. A crop might starve in a dry climate even with the best of drain irrigation: for the plant is chiefly

6. In the Skinner System of top irrigation the water is distributed through lengths of piping (usually three-quarter to one-inch pipe) which are drilled at intervals of three feet, a brass nozzle of special shape being screwed into the hole. The pipes are placed on posts three feet (or more, if desired) above the beds, and so connected at the supplying end that they can be easily twisted from side to side, so as to send the water in a fine mist wherever desired.

composed of air and moisture. It is but a very small percentage that is solid, and comes from manure or the natural earth. To prove this, it is only necessary to dry a great heap of vegetable growth (or of grass) and to burn it; a very small heap of ash will be the result, and this represents the solids which the great heap contained.

MULCHING

The value of mulching in hot weather is overlooked. Tomatoes and lettuce are two vegetables that benefit greatly by mulching. The ground can be mulched with a number of materials-old manure, leaves, straw sweepings, anything in fact that will keep the moisture in the soil by preventing the hot sun from evaporating it. The benefit in using manure as a mulch is that every watering takes the manurial salts down to the roots of the plant. Mulching can be carried out at any time; the best time is when the plants appear to be drying up and burning with the heat. Even well watered plots will benefit by being mulched; the best lettuce ever raised by the writer were grown on mulched beds.

MANURE

Manure of some kind or other is absolutely essential in most of the land surrounding Sydney. Though electrification of the seed makes the plant stronger and more able to obtain its food from air and soil, it does not do away with the necessity for manure altogether. Less manure is required by an electrified plant than by an unelectrified one.

AFTER-TREATMENT

If well-rotted stable manure is obtainable, it is to be preferred to any other; but it is not always obtainable, and some other must be substituted. Blood and bone manure is a powerful fertilizer and a good all-round manure, and the electroculturist will find that it usually answers his purpose. Nitrate of soda is particularly valuable to the vegetable grower if used in conjunction with phosphates and potash. Used as a top-dressing, only half an ounce to the square yard is required. It can be used with benefit when the plants are young, and again just when they are maturing. It can also be used as a liquid manure, applied at the rate of four ounces to a kerosene-tin full of water. Keep it off the foliage.

A common mistake is to suppose that bonedust and artificial fertilizers are as good as animal manure. All artificial manures to a certain extent impoverish the soil, and the increased production they give at the time is followed by a soil reaction. If you use one of these manures freely this year, you will of necessity use it more freely still next season; for these manures do not build up the land as does animal manure. Land that is freely manured with well-rotted stable manure will continually improve in quality, and will need less manure each season, because you are improving the quality of the land.

It may seem a peculiar statement, but the best manure is water. Soil is not absolutely necessary to the growth of a plant. We have always seen cabbages growing in soil, but in the near future they may be grown in some other element. Some plants, for instance, grow entirely in water; and potatoes have sprouted readily on a zinc plate fed by an electric current (see page 41). While realizing that

TEST BEDS AT ST. QUENTIN NURSERY.
ELECTRIFIED LUCERNE ON BOTH SIDES OF MAIN PATH.

commercial fertilizers are the only ones obtainable in certain places, we cannot allow the claim that they fill the place of animal manure. Water, it is claimed, never occurs entirely pure, because it is the most efficient solvent at present known, and therefore contains in solution some of the soil-constituents through which it has passed. Rain-water is commonly accepted as being the purest water known; but in passing through the atmosphere it absorbs many other substances, such as nitrate of ammonia, carbonic acid gas and other acids-also electricity, if it has traversed electrically-charged areas.

Professor Nobbs of Saxony has ably demonstrated that water is more essential than soil. Near his forest school could be seen, prior to the war, great trees that had been growing for years in immense glass jars of water, to which certain amounts of nitrogen and other plant-food were added once a month. No soil was necessary to produce these perfect specimens of vegetable life. At the same school experiments went to prove that for every hundred pounds of wheat gathered only one per cent. came from the soil.

The value of nitrate of soda is too often overlooked. It may be beneficially applied to almost any plant— except legumes, which have their own nitrogenous manure system. Nitrate of soda is easily taken in by the soil on account of its soluble nature. The action of the atmosphere, or a slight shower, will take it down to the roots. Top-dressing with this fertilizer is generally satisfactory, though in certain instances it will be more convenient to use it in the form of a liquid. When used in this manner it should not be allowed to touch any foliage.

The value of a compost-heap is also often overlooked by the beginner in gardening. A good idea is to have a hole dug about four feet deep and place all green stuff, leaves, &c., in it. Animal manure can be added and all classes of waste matter. The main thing is to keep the heap of refuse moist; it is therefore a good plan to allow the washing water to find its way into the pit. The heap should be covered over with a layer of soil, as this prevents the ammonia which arises during fermentation from being lost. In selecting a place for the pit a natural clay bottom is desirable; otherwise the bottom should be cemented, or a lot of the manurial value will escape. Alternate layers of earth, lime and refuse make an ideal compost. Vegetable manure obtained from the compost pit is the most valuable of manures, as it improves the texture of the soil.

Liquid manure is a valuable forcer, and is especially useful in speeding up cucumbers and rhubarb. It is made by mixing cow or sheep or fowl manure with water. If cow or sheep dung is used, about thirty pounds of dung can be used to the same number of gallons of water. If fowl droppings are used, they must be very well diluted, for otherwise the liquid will burn the plants. It is never advisable to apply liquid manure in the middle of the day, or to allow it to come in contact with the leaves.

Lime is invaluable as a sweetener to new or sour land; though it is scarcely a manure, it helps to make the natural ingredients of the soil more readily available for the roots to take up. Lime should not be applied too frequently, as too much of it impoverishes the land.

Rotation

Most gardeners realize the benefits derived from the rotation of crops. Each vegetable takes certain kinds of food from the soil, and it is not fair to any soil to draw from it year after year exactly the same constituents. Plants belonging to the one family, therefore, should not follow one another: a complete change of crop prevents exhaustion of the bed and degradation of the produce. There are cases, of course, in which one type of crop has been grown successfully year after year in the same soil: but this can only be done in deep, rich soils such as are found on the Dorrigo or some of t he South Coast flats. At Dorrigo maize has been grown on the same land without any sign of deterioration for forty years, and maize is a most exhausting crop. But that is evidence of the wonderful richness of Dorrigo land, not of the uselessness of rotation. Gross-feeding crops, such as cabbage, should on no account be grown on the same land in successive seasons.

Seedbeds

It will be found advisable to set aside certain beds in which to raise plants for the garden. In these beds the soil should be very fine; if it is very clayey, sand should be added, as young plants like friable soil. It is not necessary—or even desirable—that the seed-beds should be composed of very rich soil; seedlings can be successfully raised in plain sand. If the seed-bed soil is too rich, the young plants receive a set-back when transplanted into ordinary garden soil. Plenty of water is essential, and a very

CUCUMBER SEEDLINGS PLANTED ROUND AN INTENSE IRRIGATOR.

A METHOD ADOPTED AT WENTWORTHVILLE WITH GREAT SUCCESS FOR SECURING EARLY CUCUMBERS. LITTLE SPACE IS USED, AND WATER IS APPLIED RIGHT AT THE ROOTS. ELECTRIFIED SEED USED.

fine manure mulch will assist the young seedlings to progress. Do not sow the seed too thickly, or the result will be spindly growths. Remember, with electrified seed the speed of germination is greatly increased.

Seed-beds should have an easterly aspect. They should be well drained, absolutely free from weedgrowths and well pulverised. The soil for a depth of at least six inches should be passed through a sieve. The beds should be levelled off, and a layer of sifted rotten manure placed on them, this in turn being covered with a half-inch depth of really good soil; then they must be rolled to a proper surface and thoroughly soaked. Sow the seeds broadcast or in drills, as preferred; if broadcast, do not sow too thickly. Water the whole with a fine spray, and in the case of delicate seeds (such as cauliflowers) caver with sacking to keep cool. Remove this after the heat of the day is past. On no account allow the beds to get harsh and dry.

When plants are well up, the cover can be done away with, and a little sifted manure be layered on. Water seed-beds freely, for if the germinating seed gets at all dry in hot weather it may be a failure.

Raise your own plants always, as unelectrified plants will generally prove disappointing.

TESTING FOR GERMINATION

Not even the electric current can make dead seeds grow. You should therefore always test for germination (i.e. discover what proportion of the seeds in a packet are likely to grow) , and thus avoid sowing bad seed—of which a great deal is at present on the market. A simple method of testing is as follows: place upon a plate a piece of folded flannel damped with warm water, and on this flannel place the seeds to be tested. Stand it away in a warm place. In a f ew hours (twenty-four at most) the fertile seed will have sprouted. By counting the sprouting seeds the percentage of germination can easily be arrived at.

Some of the most successful gardeners prefer to raise their own seeds; but this needs very great care, and mistakes will be frequent until the grower has had long experience.

Transplanting

When transplanting in dry weather, it is inadvisable to apply any top water to the plants for some days. Make a hole to contain the plant, and fill it, with water until it will absorb no more. Press the plant down firmly, so that contact may be assured, and see that the top earth is thoroughly dry. Lettuce do not transplant well in dry hot weather: if it is found necessary to shift them, they should be shaded with tea-tree or other suitable protection. But in summer such salad-vegetables as lettuce do far better if planted where they are to mature; no time is then lost by throwing-back after transplanting—though the electrified plant is less affected by transplanting than those that have been raised from untreated seed.

If proper care is taken there need be no failures in transplanting. Plants from electrified seed, having a better rootlet system, get away quicker after transplanting.

A Cabbage Plant twelve weeks old, raised from Electrified Seed. Note the healthy growth and absolute freedom from aphis or other disease. Taken from six feet above.

The Root System

One of the chief benefits derived from electrifying seed is the improvement in their root system. The root is the most important part of the young plant, for a healthy root system means a sturdy plant. Though the root cannot take up solids, it carries much solid food to the plant, dissolved in the water which is carried along the little rootlets to the great roots. The tips of the roots are very sensitive, and are protected by Nature with caps, or they would be injured in their efforts to keep the plant supplied.

To demonstrate how the roots travel in search of moisture it is only necessary to mention that on the Hunter River lucerne roots have been found at a depth of forty feet. In the illustration given of a Dwarf Champion Tomato plant raised from electrified seed, it will be seen that the root system is nearly as long as the branch formation, giving the young plant better facilities for taking nourishment.

ELECTRIFIED DWARF CHAMPION TOMATO.
PLANT THREE WEEKS OLD.
NOTE VIGOROUS ROOTLET SYSTEM.

BLOOD-RED BEETS.
THE LARGER, THE PRODUCT OF ELECTRIFIED SEED, WEIGHS 2½ LBS.
THE SMALLER, FROM UNTREATED SEED, WEIGHS ¾ LB. BOTH SOWN SAME DAY.
50 DAYS MATURING. GROWN AT WENTWORTHVILLE, N.S.W.

CHAPTER IV

SPECIAL VEGETABLES

As the electroculturist is also an "intense" cultivator and quick deliverer of the goods, he must plant varieties that mature quickly. It is of little use to hustle up a slow-growing variety, when by applying hustle to the right kind of seed we get results weeks earlier. After nine years' persistent experiments in the use of electricity and quick-growing methods, certain varieties appeal to the writer as the best. Of course, any variety will mature more quickly when the seed is electrified, but naturally quick-growing varieties are made quicker still. Again, certain varieties have proved to be more susceptible to our treatment, or have given better results to the experimenter. The following suggestions may therefore be worth consideration.

In *Beet* I have found "Early Egyptian" one of the best.

Some of the very best results that we have obtained with *Cabbage* have been with "Improved St. John's Day" and "Ail Head." Try "Curly Savoy" for winter growing.

Cauliflower: special success with "Early Mammoth."

Cucumber: the best results were obtained with "White Spine", "Apple-Shaped" and "Commercial" were also successes.

Beans: without doubt the most successful variety tried was "Canadian wonder."

In *Broad Beans*, "Broad Windsor" gave the best results.

Exhaustive tests were carried out with several kinds of *Carrot*, but "Early Short Horn" always gave the best result.

The old favourite "Hollow Crown" *Parsnip* I found hard to beat.

The best *Lettuce* from an electroculturist's standpoint was "Neapolitan." The beautiful crisp hearts came quickly, and were always favourites with the buyer. The only drawback was that during the very hottest month s they scalded badly, and had to be protected on hot days. "Kew York" stood heat better.

White Turnips give great results under electroculture, and in our experience the best variety was "White Kepaul."

In *Peas* "Richard Seddon" deserves special mention, though splendid results were obtained with "Yorkshire Hero" and "American wonder," the latter giving well-filled pods more quickly than any other variety.

Radish may be termed the electroculturist's specialty, as wonderful results were obtained. Two varieties stood out above all others as consistent croppers and quick growers, namely "French Breakfast" and "Long Scarlet."

Watermelons: try "Kleckly Sweet," "Cole's Early," "McIvor Sugar" and "Cuban Queen."

In *Squashes* the "White Custard Bush" gave the best results.

Tomato-growing offers a great field for the electroculturist, as early tomatoes are always worth a deal of money. There are many splendid kinds, and it is difficult to single out a few amongst them. However, some of the best are "Matchless," "Chalk's Early," "Jewel," "Earliana," "Burwood Prize," "Mikado," "Recruit," (a splendid germinator and healthy grower) and "Golden Sunrise."

Beetroot

The growing of beet has been sadly neglected in New South Wales, though the ever-increasing demand for this salad vegetable (owing to the necessity for a suitable addition to the cooked beef and potted meats which are now largely used) makes the beet increasingly popular. Beet grows well in nearly every class of soil of good quality, but needs an abundant supply of water. If manure is added before the beet is sown, it should preferably be well-rotted stable manure, and should be dug well in. Superphosphate and guano are also recommended. Sowing can be carried out in every month from July to December, and again in February and March if suitable weather prevails. The seed should be soaked in liquid manure and nitrate of soda a few hours before sowing, but after electrification. Seed is best sown in drills, rather thinly, and covered to the depth of an inch. In weeding and chipping great care needs to be exercised, as beet, if touched ever so lightly, bleeds and is rendered useless. Some persons transplant beet. If this is done, it will need great care; if the roots are broken or bruised the crop will be worthless.

There are many good varieties, but from an electroculturist's standpoint "Egyptian" takes a lot of beating. The round or "Globe" varieties are much more popular than the elongated ones, and better sellers.

The demand for beet is not likely to diminish, especially during the hot summer months. The retail price of beetroot is never much below four-pence per bunch of four, so it pays well.

Cauliflowers

Perhaps no vegetable is more appreciated than a good cauliflower, and growers who can secure good heads find a ready market at a profitable price. If once cauliflower is checked, it is doomed; so it must be planted in a suitable bed of rich soil. A frame of hessian should be used to cover the tender plants in the early stages; but it must be removed each evening after the heat of the sun has passed, in order that the plants may not become too spindly and weak. Cauliflower is one of the vegetables which cannot be successfully crowded; when transplanted it must have sufficient room, even three feet apart each way not being too much.

Up to fifty tons of manure per acre is not too much for cauliflowers, for this vegetable does not answer to the general rules of electroculture in regard to manure. Liquid manure helps cauliflowers to head up well, and the ground between the plants should be kept well cultivated. When commencing to head, the leaves should be tied over to protect the "flower" from the sun and weather. In Mr. Herbert J. Rumsey's book, "The A.B.C. of Vegetable Growing," the latest and best methods of cauliflower-growing, adopted by the New York truck-growers, can be studied.

Celery

The variety generally preferred for "intensive" celery culture is "White Plume"; electrified seed can be sown in August or September in prepared boxes of good soil. In about a month or six weeks the celery should be ready to transplant—one or two inches apart in a small bed. Remove any spindly growth, and make the plants grow quickly by giving plenty of water and having the bed rich.

When the plants are considered large enough for the permanent bed, they should be again transplanted and placed five inches apart in rows ten inches apart. The surface of the permanent bed should be kept well stirred; we prefer, however, to have it mulched with rotted horse manure. The plants should fairly hop along; if hanging back at all, liquid manure should be freely applied. As they grow, they should form a dense mass shading each other. When nearing maturity a ten-inch board should be run around the outer edge of the bed to assist bleaching. Celery grown in this manner is just as choice as that grown in the old trench method, and it is easily seen that a great economy of space is secured. The main thing to remember about celery is that, to be of the best quality, it must not receive any check in its growth. If part of the bed is removed, the outer edge left should be boarded up to keep the celery a good colour. Before marketing, any discoloured stalks should be removed, as stained celery is not attractive. Celery prefers well-drained rich soil, and must have a plentiful supply of water.

Cucumbers

A light sandy soil suits cucumbers best. It should be made as rich as possible by the addition of plenty of well-rotted manure, which should be dug in in a circular patch about three feet across and with at least three feet from outside edge of next patch. In the centre of this circle should be placed an "intense irrigator" pot (No. 3 size, 12½ in. by 5½ in.). After electrification the seed should be sown about six inches from the pot edge, six seeds being placed round each pot at even distances. The seed should not be sown more than an inch deep. After the plants are up and making second leaves, the forcing should commence, and the pots be filled every third night with all the liquid manure (previously strained) that they will distribute. The following evening water should be freely flooded into the pots; the hose may be put into the pot, and allowed to run slowly until the soil will not absorb any more water. This treatment should be continued, and the shoots kept well pinched back to induce the cucumbers to set freely. Hundreds have been taken off a small bed by this method.

PUMPKIN, CUCUMBERS AND TOMATOES INCREASED IN SIZE OWING TO ELECTRIFICATION.

If early, some provision should be made to guard the young plants against frost, such as covers which can be removed during the day; for the cucumber, along with most other vegetables, loves the sun. When once the plants are up and start to run, they should be manured every week with liquid manure, made with either cattle dung or fowl manure (the latter is very strong, so must be greatly diluted before use). The runners soon get a move on, but must not be allowed to travel over two feet before they are pinched at the ends. Do not allow fruit to form on the main stem; it should come on the side-shoots. These side-shoots must be again pinched, and will throw out other side-shoots; stop these at the first leaf. If the situation is very windy, it may be necessary to peg down the vines to save them from damage. The cucumbers should be watched, and arranged so that one's growth will not interfere with another's. Often they set in cramped positions, and get pinched and spoilt for market, though a little attention could have turned them out good market specimens. Don't be afraid that pinching will hurt the vines; they will grow stronger than ever, and carry ten times more fruit. "White Spine" and "Commercial" cucumbers will be found good varieties. If the leaf foliage becomes too thick, it may be necessary to thin that out also. Plenty of water during growth will help them along.

Cucumbers mature in the beds at St. Quentin in a fortnight after they have set. Properly cared for, the vines will continue to bear for a long period. Some persons prefer the apple-shaped cucumber for its easily digested properties, but for market purposes "White Spine" or "Commercial" cannot be beaten.

EARLIANA TOMATOES GROWN FROM ELECTRIFIED SEED.
WEIGHING OVER A POUND APIECE. SEASON 1920-1921.

Onions

The onion is easily one of the most valuable vegetables, and enters into many dishes. It can be grown in most parts of New South Wales, but needs a well-prepared soil.

Like the turnip, it prefers land that has been manured for a cabbage crop. The best soil is a friable loam, rich in humus. It grows also in sandy soils.

Seeds should be sown in March if the best results are desired; when the seedlings are strong, transplant them in rows. Onion seed is a slow germinator; consequently the weeds usually germinate ahead of the onions, and choke the tender growths. Hand-weeding should be carried out as soon as possible, as onions above most vegetables like clean land. If the seedlings are too thick they may need thinning. One of the best varieties of onions is "Brown Spanish."

Parsnips, Carrots and White Turnips grown at St. Quentin Nursery.
Carrots matured in 50 days, white turnips in 35 days, parsnips in 60 days—half the usual time.

ELECTRIFIED AND NON-ELECTRIFIED CUCUMBERS.

THE FRUIT FROM THE TREATED SEED WEIGHED 6 ¾ LBS, THE OTHERS 2½ LBS. BOTH BEDS RECEIVED SAME GENERAL TREATMENT. TEST AT ST. QUENTIN NURSERY.

Parsnips

Perhaps no other vegetable in the garden fails as often as the parsnip. The seed is the shyest of germinators, and if the bed gets at all dry during the germinating period failure results. The electrification of the seed has been found to be most beneficial, as up to 92 per cent. germination has been secured. Parsnip seed is a bad keeper, and should only be electrified just prior to sowing. It prefers soil of a sandy nature, and fairly deep. No fresh manure should be used. If the bed is a small one, the covering of it with hessian during the first few days will materially assist the plant to establish itself. The main crop is sown early in the spring or late in the winter. To ensure parsnips, the bed should be kept absolutely free from weeds and a plentiful supply of water be assured. Parsnips ordinarily take 125 days to mature; with electrification the time is much shortened— they have been produced in 100 days by this method.

If new manure is placed in carrot, parsnip or other root-crop beds, the product will fork and be useless for the market. Place the parsnips and carrots in the deeper positions if the land is on the average shallow.

CUCUMBERS ON THE VINES (WHITE SPINE AND CTOMMERCIAL VARIETIES), SHOWING HOW ELECTRIFICATION OF SEED CAUSES MORE FRUIT TO SET. MANY WEIGHED UP TO 2½ LBS. EACH.

Rhubarb (variety Giant Red) that has received electric simulation. Note the healthy growth. The electrified root gave three cuttings, the others two.

Rhubarb

A method of rhubarb culture that appeals to the electroculturist is practised by many up-to-date growers. A variety that does not cease growing during the winter is chosen as most suitable, and the ground is very heavily manured in advance. A deep sandy soil is ideal.

The electrified seed is sown in February, or even March, in drills two feet apart, and the plants are thinned out to eighteen inches apart in the rows. Water is applied every week, and the bed thoroughly soaked. Plenty of liquid manure (preferably cow manure) will make the rhubarb leap ahead, and a good crop of long stalks can be maintained well into the early summer. Once the plants are played out, they should be dug up, and fresh seed sown for the coming season.

Strawberries

The cultivation of the strawberry will be found most profitable if the soil is suitable, e.g. a sandy loam. Spread rich compost in between the plants, and when they start to make headway cut the runners. Give plenty of water during early stages of growth, and use liquid manure and water alternately. When the fruit commences to colour, stop using liquid manure altogether, and give very little water. Too much water at that time will spoil the flavour of the fruit. "Creswell's Seedling" gives good results in most places, but "Marguerite" is of superior flavour. When birds are troublesome, frames of bird-proof wire should be placed over the beds whilst fruiting.

Strawberry plants may be purchased of most nurserymen, and the beds can be increased by planting runners. As the highest price is paid for all early vegetables, it should be the electroculturist's chief aim to produce as much early crop as possible. This is where a suitable hot-house comes in.

Good boxes for sowing seed under cover can be made from butter-boxes cut in three and bored at the bottom for drainage. A little broken stone and leaf at the bottom will be of use. All plants that have been raised under cover should be hardened off, by placing in the open air for a few hours a day, before they are finally transferred to their permanent position. The hardiest plants are those raised in the open, but it is not always practicable to raise them without covers.

Tomatoes

The present season (1921) has been one of the worst on record from a tomato-grower's viewpoint. Wilt, red spider and leaf disease affected the majority of crops, and one grower lost every tomato planted. It is noteworthy that there was no exhibit of tomatoes at Gosford Show, owing to the failure of the crop there—and Gosford is one of the best tomato-growing districts in the State. But tomatoes from seed electrified at St. Quentin were planted out both in the experimental area and on Mr. Bugler's and other neighbouring farms, and all remained surprisingly healthy. Out of twelve hundred planted by Mr. Bugler not more than a dozen were affected at all. These tomatoes were put in without any manure whatsoever, and in rough broken ground, from which the conch grass had not been removed; so they had none of the advantages of cultivation. The soil round the plants was in many instances as hard as macadam, and they were nearly all left unpruned and unstaked; yet they bore a very heavy crop of sound tomatoes. The varieties were "Sparks' Earliana," "Dwarf Champion" and "Burwood Prize" mostly, with some "Matchless."

The illustration opposite shows a staked plant that had fifty tomatoes on it, averaging half a pound each in weight. This was one of many, and its produce at the market rate of the time would be 18s. 6d. As over nine thousand plants can be accommodated on an acre of ground, this would give us about £7,000 per acre return. Allowing the crop to be half as good on the average, it would mean at least £3,500 per acre return.

The tomatoes planted out on the experimental area are all doing well, the fruit setting in clusters of up to a dozen in a bunch. Mr. Bugler's heaviest bearers were "Burwood Prize" and "Earliana." The heaviest fruiters on the area were "Duke of York" and "Dwarf Champion."

Buyers report that the tomatoes transplanted to their areas from St. Quentin are all doing well. Mr. Walters, of Wentworthville has 450, mostly "Matchless," making splendid growth. Practically every tomato was clean.

Very little manure of any kind was used on the test lots, though a little nitrate of soda was placed on the surface of the "Dwarf Champion" beds, as the experimenter was assured by local growers that this particular tomato was a failure in the Parramatta district. From results at St. Quentin,

DUKE OF YORK TOMATOES, GROWN FROM ELECTRIFIED SEED
AT ST. QUENTIN NURSERY DURING 1920-21.
NOTE FREEDOM FROM SPOT.

however, it would seem to be one of the best. It is a sturdy grower, resists heavy winds better than most tomatoes, and requires very little attention. The heavy bearing in the test beds shows that close growing is not detrimental to heavy cropping.

White Turnips

To be of any value white turnips must be grown quickly, at any rate in not over forty days. Ground that has been heavily manured is best for the white turnip. No new manure should be used—if any manure is applied, it must be well-rotted. Bonedust is a good manure for this vegetable; but over-manuring with bonedust has a tendency to impoverish the soil, and not more than two cwt. per acre should be used. Turnips, like cauliflowers, will not stand a check, and a plentiful supply of water is an absolute necessity. Without plenty of water during the hot dry months they will grow staggy and hot. Turnips pay better if sown in rows, as they can be weeded between and kept clean. The land should be finely worked if the best results are desired. "White Nepaul" is the electroculturist's favourite.

BURWOOD PRIZE TOMATOES, FROM ELECTRIFIED SEED.
PLANT HAS OVER 50 TOMATOES, WHOSE AVERAGE WEIGHT IS HALF A POUND.
THESE TOMATOES RECEIVED NO CULTIVATION OR WATERING, AND THE LAND IS HARD
AS MACADAM. ALL THE VINES ARE EXTREMELY HEALTHY. GROWN BY W. BUGLER.

Table of Monthly Sowings

January sowings may be made of cabbage, cauliflower and tomatoes in prepared beds; in the open, beans, beet, carrot, early cucumbers, lettuce, parsnip, kohl rabi, early varieties of peas, radish, turnip, and silver beet.

February: in prepared beds, cabbage and celery; in the open, French and butter beans, early varieties of beet, early carrot, kohl rabi, lettuce, parsnip, parsley, early varieties of peas, potatoes, swede and white turnips.

March: broad beans, early cabbage, early carrot, kohl rabi. lettuce, early onions, tree onions, radish, shallots (if available), swede and white turnip, and early beet. If the district is not troubled with heavy frosts, sowings can also be made of early beet, French beans, parsnips, and dwarf varieties of peas.

April: broad beans, early cabbage, early carrots, kohl rabi, leek, lettuce, onions, parsley, parsnip, tree onions, radish, swede and white turnips and prickly spinach. Also a small sowing of peas if the weather is not too cold.

May: broad beans, onions, radish, beet, early cabbage, early carrots, kohl rabi, parsley, tomatoes (under cover), lettuce, parsnip, early varieties of peas, white turnips.

June: broad beans, early cabbage, early carrots, lettuce, parsley, dwarf peas (early sorts), potatoes, prickly spinach, radish, white turnip; tomatoes under glass or in a hot-house.

SPECIAL VEGETABLES

July: broad beans, early beet, St. John's Day cabbage, early carrots, leek, lettuce, early onions, parsnips, parsley, peas (main crop), potatoes, radish, spinach, turnips.

August: asparagus seed, artichoke, beet (all sorts), cabbage, carrot, leek, lettuce, early onions, peas, parsnips, parsley, radish, salsify, turnip.

September: French beans, beet, cabbage, artichoke, carrot, celery, chili, cucumbers, leek, lettuce, melons and squashes, pumpkins, onions, parsley, parsnip, peas (early sorts), potatoes, radish, a little rhubarb, tomato, turnips.

October: beet, French and butter beans, cabbage, early carrots, cucumber, leek, lettuce, melons, watermelons, rockmelons, salad onions, parsley, parsnips, peas (early sorts), pumpkin, radish, silver beet, squash, tomato, turnip.

November: Beet (early sorts), French and butter and climber beans, cabbage, carrots, cucumber, lettuce, melons of all kinds, squashes, marrows and pumpkins, radish, silver beet, sweet potatoes (plants of), tomato, white and swede turnips.

December: beet, all kinds of beans except broad, cabbage, cauliflower, cucumber (early sorts), lettuce, radish, silver beet, squashes, swedes (small sowing), sweet potato (plants), tomato, turnips.

A Bed of Silver Beet with stalks up to three feet long.

CHAPTER V

MISCELLANEOUS HINTS

Economizing Ground

ELECTROCULTURE goes hand in hand with "intense" cultivation and implies both quick and close growth. Rows three feet apart disappear, for the electroculturist wants to produce as much as possible to the inch. Beds, therefore, should not be large. On a bed nine feet by twelve, sixty fine-hearted cabbages were matured in twelve weeks at St. Quentin—cabbages that sold later on for eightpence to a shilling apiece. Moreover, the small bed is more easily weeded by hand; and hand-weeding is the only satisfactory method when you are dealing with small plants that have been sown in the permanent beds and are not to be transplanted. Similarly, lettuce grown close together are always crisp and tender, besides being self-blanching.

Oblong beds are as a rule more easily worked than square ones. The strips of land taken up with paths are not wasted; paths at frequent intervals save the beds from unnecessary trampling.

Vegetables such as beans may have lettuce interposed between the rows. This method may also be used with carrots and parsnips, so that a crop of lettuce can be taken off the beds whilst the slower parsnip or carrot is maturing.

The Value of Trellises

Too much emphasis can not be placed on the value of trellises in economizing ground. For instance, if epicure beans are planted at the foot of proper trellises, it will not be necessary to give much garden room to dwarf French beans, and more of the salad vegetables—which are undoubtedly money-makers—can be grown. Certain varieties of tomato (such as "Walker's Recruit") will give splendid results on trellises with an easterly aspect. Many gardeners grow most of their tomato crop in this manner. Fruit grown on trellises does not suffer from rot in wet weather, and gets the full benefit of the morning and evening sun, and plenty of fresh air. Trellises need not be over six feet high; that height will suit most climbers, and the product is more readily picked than if the trellis were higher.

A trellis can be made of hardwood uprights and green timber, with battens at top and bottom, and covered with netting wire. Such trellises are easily constructed, act as shelters and breakwinds, and form an ornamental backing to the garden. The stretches between the posts should not be too great; otherwise the weight of passion-fruit or other crops may cause the trellis to sag. Two trellises, one on each side of the path, can be joined overhead with battens to form a neat archway.

WHITE NEPAUL TURNIPS, FROM ELECTRIFIED SEED.
MATURED IN THIRTY-TWO DAYS. THE TREATED BED GAVE A THIRD MORE
GERMINATION THAN THE UNTREATED CHECK-BED, AND THE TURNIPS
WERE READY NEARLY A FORTNIGHT EARLIER.

Piping and Hose

A little extra outlay at the beginning in piping will be repaid. In a garden sixty feet by a hundred and twenty, if a pipe-line is laid up the middle, it will be advisable to have two taps forty feet apart and forty feet from each end. With a forty-foot hose every part of the garden is under command. In buying a hose see that it is pliable, as stiff hose bursts more easily, and is heavy to carry round. Avoid long lengths of hose; you will find forty feet long enough to move about. Have more standpipes and less hose. If you are not in a position to instal a sprinkler system, or only have a plot for home use, one of the revolving sprinklers that can be purchased for a few shillings will save you a deal of work, and do the watering better than it would be done with the hose direct. The writer prefers the "Marsh Rainmaker," made by a Melbourne firm, to any other type he has inspected. Sprinklers of this type have been in use at St. Quentin since its inception, and have given every satisfaction. The throw is wide and adjustable, and the sprinkler is simple and easily overhauled.

A little contrivance to hold the hose off the garden during shifting operations is very necessary. It can be made with a piece of three-by-two hardwood. Take a four-foot length, sharpen one end, and saw the other end off straight. Take a second piece about eight inches long, and saw it into a U shape, leaving the bottom of the letter square and thick enough to give strength. In this bottom bore a hole to take a four-inch bolt easily. In the arms of the U make holes four inches from the top to take a smaller bolt; run the bolt through one of these holes, then through a wooden roller (which should move easily on the bolt), and then through the hole in the other arm of the U. Nut up firmly, and the wooden roller will revolve. Take it out again, and drive the four-inch bolt through the hole in the bottom; this hole must hold the bolt easily, not tightly. Make a smaller hole in the top of the four-foot piece of hardwood; drive the bolt tightly home in this, and you will have a revolving short top piece. Screw up the wooden roller again, and drive the whole into the ground to a depth of eighteen inches or two feet. You now have an attachment over which the hose will pull readily, and which will keep it off the intervening beds. One of these hose-raisers should be placed opposite each tap on either side of the garden at a convenient distance.

Another useful piece of apparatus is a garden roller. A piece of heavy round timber, such as a portion of the bole of a small tree, with the aid of a few pieces of batten and a couple of bolts, makes an excellent and easily constructed roller.

For planting small seeds in rows where it is not desirable to cover them too deeply, a handy seed-sower can be made by driving pegs of wood through a bit of batten at a distance corresponding to that between the rows. These pegs should be slightly sharpened. By attaching a handle, this contrivance can be drawn over the prepared bed, and perfectly level and straight drills of any desired depth may be easily made.

Canadian Wonder French Beans growing on poor soil in couch-grass land that has only been slightly chipped with mattock. A striking proof of the vitality of electrified seed. A similar bed gave record crop. The beans break up land and make after-cultivation easier.

Application of Current to Growing Crops

Undoubtedly the application of a medium current to growing plants has a forcing effect. To prepare the bed for an underground current, after thoroughly digging it, sink in it copper wire at a depth of from three to six inches. The wire should be worked into a series of N's and held in position at each turn by pegs, the aim being to make the current traverse as much of the ground as possible. The terminals should be taken to an insulator, and the ends lightly attached, so that they can be connected with the pales of the electrifying apparatus.

The writer has found that the introduction of another current into the leaf foliage at the same time has a good effect. Overhead wiring should be somewhat in the form of a woven mattress; the turns at each side will be held by proper cup insulators, mounted on poles set in the ground. The wires should not be more than six inches above the plants when growing. If a crop such as oats or peas is wired, the wires must be so fixed that they can be raised easily as the crop grows. Plants take the current better after a good sprinkling or a fall of rain. If the current is applied weakly at first, and gradually increased, a greater benefit will be noted.

Wiring beds, however, on anything like a large scale can not be contemplated by the average person, as wire is too expensive; and, since the same result can be obtained by the electrification of seed, it is unnecessary.

TREATING A SICK CABBAGE WITH PRIMITIVE APPARATUS, CONSTRUCTED FROM A FISHING-ROD, GLASS PHIAL AND ZINC PLATE. THE DISC CAN BE LOWERED TO TOUCH THE PLANT. THE ELECTRODE CARRIES THE GROUND CURRENT. THE TREATED CABBAGE WAS FIRST TO MATURE IN BED.

Weeds, Pests and Remedies

Electrification of seed is a great help in the struggle with weeds. The gardener's life is largely a hand-to-hand combat with weeds; they start on at least even terms with the crop, benefit equally with it from any manure applied to the beds, and can be always depended on to have a thoroughly healthy constitution. But electrified seed has been given a start to which the weeds—if not encouraged by the gardener's neglect—can never catch up, and will play their own game against them, overgrowing and choking them out.

The ordinary red ants, that build a gravel fort in the paddock, will prove a nuisance to the gardener by placing aphides on the beans or other vegetables. They carry these insects from one place to another, and depend on them for a sweet substance that exudes from their body, indeed, the aphis is sometimes called the ant's cow. When once the ants decide to form a feeding-ground for their cows—say on the broad bean patch—the electroculturist should get a move on. Here the battery will again come in handy. The current, strongly administered, will kill the aphis, and a charged wire across the ants' regular road will make them think the locality unhealthy.

On the other hand, many garden-owners look on every living thing as an enemy, and straightway set out to kill their best friends. Some birds do damage, but do a deal of good as well. The little silver-eye, for instance, is keen on aphis, and will be seen amongst the cabbages. The swallow, the pee-weet, and even the sparrow do good. The lady-bird beetle accounts for thousands of aphides, and its larva lives on them. (Often the lady-bird beetle is confused with the pumpkin beetle, though in reality there is not a great deal of similarity. The pumpkin beetle is elongated, not round like the lady-bird.). Bees and butterflies, in common with some insects, insure that your garden will produce fruit. The frog is the gardener's friend also. All sorts of birds flock to the garden, because the green growths encourage insect life, and on these insects the birds feed. The ordinary long worm works after hours, ploughing the soil and fining it down. The small bacteria in the soil, too, are continually working on the electroculturist's behalf. In France heavy shelling with high explosives, and poisonous gases, failed to destroy these friends of the tiller of the soil. One of the worst enemies of the garden is the wire-worm; as an antidote linseed cake can be recommended—if placed in small pieces about the garden, it kills all the wire-worms that dine off it.

Growing without Gardens

Those who have only asphalted yards, or mere verandahs, can yet cultivate a fair variety of vegetables if the seed is previously electrified; electrification makes the plants hardy, and they require very little soil if an abundance of water is given them. To raise cucumbers, for instance, by the Bennett method, first secure a butter-box. Renail or screw in the bottom firmly, and bore four holes to provide drainage. Put in some soil, then an Intense Irrigator in the centre, and surround it with earth and manure. Plant round it four seeds that have been electrified, and administer through the pot any commercial fertilizer—such as nitrate of soda dissolved in water, about an ounce to three gallons. The box can be set on a verandah, or any convenient ledge where it will get the sun (easterly preferred). A wire frame or trellis will carry the vines and fruit, which will mature rapidly. The number of plants should be reduced to two by pulling out the weakest. Cucumbers grown in this manner bear surprisingly. The best varieties for box culture are "White Spine" and "Apple-shaped."

Tomatoes such as "Walker's Recruit" can be grown successfully in the same manner by persons who have no yard room. Good radishes can be raised in a little window-box about six inches deep. Plenty of water is essential, if the radishes are to be crisp and tender. The best variety for box-growing is "French Breakfast." Electrified radishes are ready for the table weeks before untreated ones. Radishes will grow thickly, and a great quantity can be garnered from a small box. Another box, sown just before the first is emptied out, will give a succession for nine months in the year.

Spinach will grow well in a shallow box with plenty of water. So can beetroot ("Egyptian"); lettuce ("Iceberg"); beans—epicure and other runners, by training on lattice or netting; tomatoes ("Egg-shape" or "Dwarf Champion"); watermelons, such as "Ice Cream"; radishes ("French Breakfast" and turnip varieties); spinach, and all classes of herbs and watercress.

Butter-boxes cut in halves make good boxes for herbs, and whole butter-boxes for cucumbers, melons and climbing beans. Radishes will only require shallow boxes with a larger surface. The soil should be fairly rich, and plenty of water is essential. Place the boxes in an easterly aspect.

Strawberries can be grown successfully in a cask. Secure a small cask, and bore a number of holes, an inch in diameter, round the side in rows a foot apart horizontally and eighteen inches vertically. Secure healthy plants, and electrify them by placing on a plate as for potato-sets. Fill the cask to just below the first row of holes with soil and well-rotted manure; push the roots through from the outside, and fill in with soil again to just below the next row. Continue this until the cask is full to within an inch of the top. Natural manures can be administered at any time in a liquid state, but if any commercial fertilizer is added it may be used just after the first fruit sets. Only plain water must be used when the fruit is ripening. Runners should be snipped off.

Specialization

This is essentially the age of the specialist, and electroculturists will find that they can make more money by specializing in two or three standard lines. For family use, of course, practically every vegetable may be grown; but for supply purposes the grower will be well-advised to confine his interest to a few. If the area to be worked is very small, he will find that salad vegetables offer the best inducement. For these there is always a ready market; they take up little room, and under the Bennett system mature quickly. Lettuce will be found especially profitable during the hot summer months—but some means of providing shade is essential, if tender kinds are to be raised. Hessian shelters that can be drawn over the heads will be suitable. Only the severe sun hurts the heads, so it will only be necessary to use the shades during the hot hours. By thus limiting their use the plants will be kept hardy, and not become weedy, as they might if raised under a permanent shelter. If a handy wall provides shade during the hottest hours, the lettuce bed should be placed beside it.

Cabbages will not on the whole be found to pay the electroculturist; there are scores of large growers who at certain periods send in truck-loads of them, grown under field conditions, and glut the market. But the growing of salad vegetables needs much care and plentiful supplies of water, so that there will be no competition in this line from field growers. Peas and beans suffer under the same disadvantages as cabbages.

Lately the ruling prices obtained for bunch vegetables (such as carrots and parsnips) makes these crops extremely profitable, and even Chinese growers are giving them more attention. Root crops are mostly very healthy, and no loss from disease can be feared.

A Larger Sphere

The experiments at St. Quentin have been almost entirely confined hitherto to vegetables: but the method may be applied to flowers and crops with similar benefit. This side of the problem has been already studied in Great Britain, where at Pontypridd in Wales oat crops have been materially increased, and at Godmanstone in Dorset much experimenting has been done with wheat, oats and barley. In 1918, electrified seed of this description gave on an average over 30 per cent. increased yield in bushels per acre; the grain produced was of a better quality, and the straw longer and stouter. This is the "larger sphere" which the writer is now contemplating. Imagine how much such an increase in wheat-yield would mean to New South Wales! The seed in England was soaked in an electrolytic solution contained in a large wooden vat holding ten to twenty sacks of grain, and the current administered by means of electrodes (usually iron plates) at each end of the vat; the current was taken from the ordinary town supply, regulated by a rheostat. Now most of the wheat sown in New South Wales is "pickled" in bluestone before sowing, a bag being immersed for some hours in a cask containing bluestone solution. By placing sheets of metal at each side of the cask, and connecting up a current of proper strength, electrification of the seed could be carried out without waste of time while the pickling was going on. The proper time, strength of solution, strength of current, and duration of the subsequent drying period have not yet been definitely fixed by the English experiments, and must be independently determined here; this will take much careful and assiduous experimenting, and the author hopes to devote himself to this work in the immediate future.

ICEBERG LETTUCE FROM ELECTRIFIED SEED.
THESE LETTUCES HAD VERY LITTLE WATER, AND WERE PRACTICALLY DRY-FARMED.

Growing Cucumbers where there is no yard space, on a verandah in a butter-box with netting. (A side window with a little frame will answer.) The perforated pot (Intense Irrigator) is used to supply water and fertiliser.

CHAPTER VI

THE TESTS AT ST. QUENTIN NURSERY

Turning a couch-grass paddock into a nursery within six months was something of an undertaking, but it was successfully carried out. The first operation was to fell standing timber and get the area ploughed. Then water had to be installed, and some kind of fence erected to keep off cattle. All the land was then prong-hoed by hand, and the couch-grass gathered into heaps and burnt, the ashes being disseminated over the land. Up the side of the centre path lucerne was planted, a portion of the seed being electrified and the rest not (cf. p. 46). Beds were laid out on a plan that gave visitors as good a chance as possible of inspecting developments. All test-beds, treated and untreated, were numbered and entered in a book, with a description of special manure or treatment applied (if any), date of sowing, variety, &c. They were regularly checked by comparison and the differences noted in writing. When water was applied, it was distributed as evenly as possible between the beds.

SEEDING LETTUCE THAT HAS BEEN ELECTRICALLY TREATED. VARIETY, NEW YORK, 1920-21.

Trial boxes of various seeds were also sown, treated and untreated seeds coming from the same packet. The quantities were checked and the same amount of fertilizer placed in each box; and the unelectrified seed was soaked in the same manner and for the same time as the electrified. The boxes were placed side by side and the same quantity of water supplied to each.

The results of these tests were most convincing. Some extremely good percentages of germination were secured for tomatoes, as high as a hundred per cent, being obtained with "Walker's Recruit" and "Chalks' Jewel." The seedlings differed from

the start, the electrified ones coming up healthy and making most vigorous growth. In a few weeks the electrified ones were all ready for transplanting, whilst the others had hardly got a fair start. The treated seedlings of "Walker's Recruit" were transplanted nearly four weeks ahead of the others. In the test-beds seed was set either row for row or bed for bed, and a keen check kept on the growth. In nearly every instance the seed in the treated lots germinated hours ahead of the untreated, and peas were picked from the bed sown with electrified seed a fortnight earlier than from the check bed. As it was contended by the experimenter that electrification lessened the need for manure, very little was applied—what there was in the shape of a little bonedust and cow manure. Water was supplied to the beds steadily by means of revolving sprinklers attached to two standpipes.

The soil at St. Quentin is on the light side, and inclined to set very hard on the surface when exposed to heat after watering. It was therefore found necessary to keep the surface continually stirred.

No. 1 Bed. Giant Red Rhubarb. Two roots were untreated and one treated. The treated one gave four pullings to date, the others two only.

No. 2 Test. Yorkshire Hero Peas. Those electrified were pulled just two weeks before a check bed, and were out of the ground two weeks sooner, thus making room for other crops. The electrified bed gave a third more peas.

No. 3. Neapolitan Lettuce. This was a most satisfactory test. The lettuce started to mature in three weeks, and the bed was disposed of within five weeks, and replanted. In the check bed only two-thirds of the plants hearted, and they were in the ground three weeks longer.

No. 4. Drumhead Lettuce. The results in this case were also good, but not as pronounced as with the Neapolitan variety. The check bed was slightly longer in maturing, and the hearts were not as large.

No. 5. Hollow Crown Parsnip. The bed has just been finally pulled. Every root was choice, and they were eagerly snapped up by local shops at good prices. (Parsnips were for the most part dear and inferior this season).

No. 6. French Breakfast Radish. These were marvellously rapid in maturing, taking only fifteen days. The radishes were fine, large and tender. Those in the check bed took double the time to mature, and were hot and coarse.

No. 7. (Bed planted on French double-cropping system). Lettuce and Short Horn Carrot. The lettuce (Neapolitan) were sown in rows, and the carrots broadcast. The lettuce were particularly crisp and white in heart. Two crops were taken off. The carrots were of good quality, and sold well at a paying price.

No. 8. (French double-cropping method). Lettuce and Radish, the lettuce in rows and radish broadcast. This experiment was not a success.

No. 9. Carrots, "Intermediate" variety. These made good growth and eclipsed the check bed.

No. 10. Hunter River Lucerne. This was broadcast. Heavy rain beat it flat as it came up, and the hot sun burn it out.

No. 11. Special Test of Yorkshire Hero Peas. Various stimulants were used, the first row being given horse manure, the second cow manure, and the third no manure at all: but the seeds in this bed had been electrified. The third bed was responsible for three-quarters of the total crop.

No. 12. Small beds of "St. John's Day" Cabbage. The electrified bed matured, and the plants were fit for transplanting, a week sooner than those in the check bed. When planted out, the treated ones gave a third more heads of marketable size.

No. 13. Egyptian Turnip-rooted Beet. The seed here was not electrified, but was soaked in liquid manure. The plants made good growth. A check bed of electrified seed was soaked only in water, and then electrified. It produced a market crop just as early as the other, and the beets were of a superior quality.

Checks were also sown of Broad Beans, "American Wonder" and "Richard Seddon" Peas, and "Sure Head" Cabbage; and in nearly every instance the treated beds or rows beat the untreated ones.

From Cucumbers we had the most pronounced results. The varieties sown were "White Spine,"

"Apple Shaped," "Long Green," and "Commercial." All did well, but "Commercial" and "White Spine" gave the best results under treatment. Some reached 2½ lbs in weight. (Most of this seed was specially obtained from a seed farm on the Dorrigo.) Though planted later than others in the same locality, St. Quentin cucumbers were the first on the market, and up to ninepence each was realized for them.

Water and Rock Melons, Peanuts and Potatoes were also treated; but the value of experiments in connection with peanuts and potatoes was nullified owing to heavy flood rain on the area, which destroyed the check.

Various other trials were carried out, including one with Hunter River Lucerne in rows. The portion from electrified seed gave four cuttings in seven months, the untreated portion two.

These were the main trials; and, taking into consideration the fact that the land is poor and sufficient manure was unobtainable, the results are looked upon as eminently satisfactory.

The difference between treated and untreated beds was at once remarked by every visitor—and during the first six months there were scores. The return from beds treated was so convincing that all new beds are now sown with electrified seeds; only in an occasional instance, where tests are being still carried out, is there untreated seed in the beds.

In beds used for commercial purposes the results obtained with white turnips and radishes were most pronounced. White turnips ordinarily take about sixty days to reach maturity, but the St. Quentin beds arrived at that stage in just thirty-five.

TOMATO SEEDLINGS, SHOWING STURDY GROWTH FROM ELECTRIFIED SEED.

A CORNER OF THE ELECTRIFIED MELON PATCH, ST. QUENTIN, WENTWORTHVILLE.

The quality of the turnips was far superior to those taken from beds where the seed was untreated, and they had no tendency to pithiness. The radishes, which matured in from fourteen to nineteen days, were the finest "French Breakfast" radishes ever grown in the Parramatta district, and sold at twice market price. Beetroot matured in less than the usual time, and was of excellent quality. As it was the desire of the proprietress to save as much seed as possible, much of the best was allowed to stand for seed purposes.

After a visit to the plots. Mr. Herbert J. Rumsey, the well known seedsman and grower (as well as lecturer) wrote in the *Farmer and Settler*: "The beds are singularly free from aphis and other pests which usually infest the vegetable garden-especially where there is not an efficient irrigation system."

When selling the produce, careful checks were made and the return from the various beds separately noted. From exhaustive tests in measured beds it was decided that it was possible under the Bennett system to make up to £700 per acre per annum growing salad vegetables for the market.

Some months ago the experimenter forwarded a small parcel of electrified seed ("Long Scarlet Radish") to Messrs. H. Short and Son, Seed Specialists, of the Dorrigo, for a test plot. Messrs. Short have from time to time conducted experimental plots at Dorrigo for the Department of Agriculture, and it was certain that they would give the seed a fair test. Their report, now to hand, is as follows:—"I carried out a thorough experiment with the radish seed sent, planting equal quantities of it and of untreated seed in two beds, with a dressing of superphosphate at the rate of approximately 3 cwt. per acre. The bed with electrified seed was up one day before the other, and the radishes were a little superior to those from the untreated seed, although both beds were of good quality. *The electrified remained crisp longer than the others.* I got crisp radishes from this bed for one month, and plenty of them were ten inches in length. I kept a record of germination and growing times, but unfortunately mislaid it whilst writing. They did not, however, grow in the time (15 days[7]) you grew them in. I consider the result satisfactory, as I submitted the bed to numbers of people who visited the area, not telling them which was the electrified lot, and every person singled out the electrified bed as superior."

---

[7]. The time mentioned, 15 days, did not relate to Long Scarlet Radish, but to French Breakfast, a very different type which under electroculture matures more quickly.

BUNCH OF RADISHES WEIGHING 4¾ LBS; VARIETY LONG SCARLET.
SEED ELECTRIFIED. GROWN IN POOR SOIL WITHOUT MANURE.

Tomatoes from St. Quentin ready for market; 22/- per bushel case was received for these, doubling the top price of other growers.

APPENDIX I

EVIDENCE FROM DUNTROON
(CONTRIBUTED BY W. RAYNER HEBBLEWHITE)

About 1915 a privet hedge was planted down the side entrance path and along the front fence of the quarters occupied by the Physics Laboratory Assistant at Duntroon Royal Military College. About November, 1918, two blue-gums six to nine inches in height were planted close to the side entrance hedge. Shortly afterwards a pole was erected in the centre of this hedge to carry the aerials for wireless operations. The aerials ran directly over the front end of the hedge to another pole in the field opposite.

Up to the time of the erection of the aerials growth was uniform throughout the hedge. The wireless aerials were removed after six months' use. From the time of their erection onward, continuing after their removal, the growth of the hedge under the wires was much greater than that of the part beyond the wireless pole, a sharp distinction being discernible exactly at the pole. In 1920 the affected part of the hedge had grown to a height of about nine feet, whereas the part not under the wires and behind the pole was at no part up to three feet. The high part had to be clipped back for the sake of uniformity in appearance with the front hedge, and the distinction is not now marked. An interesting feature is the growth of the hedge along the front fence. From a knowledge of electrical laws, and from the sharpness of the point of demarcation in the side hedge, one would assume that the influence exerted by the wires is in a radial direction only. On this assumption it would be expected that the front fence hedge—which is at right angles to the wires—would be affected throughout, but to a degree proportional to the distance from the wires. This is exactly what has occurred. Near the junction with the side hedge the growth is of the same remarkable order, but the height dwindles down to the farther end, where the growth has not yet been sufficient to bring the hedge up to the level adopted along the front for clipping.

Two blue-gum trees offer further evidence of the effect of the high frequency currents in the aerials. One planted near the rear portion of the hedge along the side path did not come within the influence from the wires. In 2¼ years it developed from about six inches in height to two feet six inches, being in appearance a normal young tree. The other, however, which was almost under the wires, showed truly remarkable activity. In the same time it grew from a seedling of about six inches to a fine spreading tree fourteen feet in height—a picture of vigorous good health. Such a degree of growth is in itself rather remarkable in this climate, apart from the comparison with the smaller tree. (See photograph on next page.)

As the wireless plant was not in continuous operation during the time of its erection at this site, it is suggested that, in addition to the effect from the current by the wireless plant, the aerials collected a good deal of electricity from atmospheric disturbances.

YOUNG GUMS AT DUNTROON.

APPENDIX II

RECIPES

As one of the main objects of this book is to assist the working man to reduce the high cost of living, and to make him independent of the hordes of middlemen who fatten on him, it does not seem amiss to include a few recipes, the main ingredients of which the electroculturist can pick from his own garden. The writer contends that it is possible to feed a man and his family almost entirely from that garden, and by degrees to dispense with meat diet altogether. No radical change from meat to vegetables should be made, but a gradual lessening of the quantity of meat and increase in the vegetable diet will prove that this is feasible. When once the vegetable-eating habit replaces the meat-eating one, the working man will be healthier, happier and wealthier.

SCOTCH STEW

Three tablespoons pearl barley, ½ small cabbage or 1 small lettuce shredded, ½ head celery, 3 onions, 2 carrots, 1 turnip, all cut into large pieces, to which may be added other vegetables in season, salt and pepper to taste.

Method: Blanch the pearl barley, cover it with cold water, and simmer gently for half an hour. Place all the vegetables, except the cabbage or lettuce, in a stewpan, add the pearl barley and the water in which it was cooked, together with boiling water to barely cover the whole, and season to taste. Boil gently until the vegetables are nearly done, then add the shredded cabbage or lettuce. Cook for 10 minutes longer and serve.

POTTED BEANS

Half-pint Haricot beans, 2oz. breadcrumbs, 2oz. strong cheese (grated), 2oz. butter, cayenne, pepper and salt, nutmeg to taste.

Method: Bake the beans in a slow oven, pound them in a mortar, adding gradually the other ingredients. Press the mixture in to pots, and run a little butter over the top if it is to keep many days. Potted beans make very good sandwiches with bread and butter. Store in a cool dry place, as all kinds of beans quickly ferment.

BAKED TOMATOES AND BREADCRUMBS

Butter a pie dish, sprinkle in a few breadcrumbs, then a layer of tomatoes; season with salt and a little sugar, and minced onion if liked. Continue in this way until the dish is filled. Have breadcrumbs for the top, and place over that some small pieces of butter. Bake for about half an hour.

VEGETABLE CROQUETTES

Rub boiled potatoes through a sieve, add cooked vegetables, butter, chopped parsley and yoke of an egg, season with salt and pepper; form into balls, roll in flour, pepper, salt, white of an egg and some breadcrumbs; fry golden brown in fat and drain. Ingredients: 1 lb boiled potatoes, 1 teaspoonful of butter, 1 egg, 1 tablespoonful of flour, pepper, salt, 1 cup of white breadcrumbs, cold vegetables such as peas or cauliflower mixed.

Tomato Soup

Wash tomatoes, peel onions, cut up tomatoes and anions, put them in a saucepan, cover with water, add salt, pepper and sugar, boil till tender, pass through sieve, heat the butter, add flour and stir well, add milk, soda and tomato pulp, stir well, boil two minutes and serve. Ingredients: 8 large tomatoes, 2 onions, 1 tablespoonful butter, 1 tablespoonful of flour, 1 pint of milk, 4 pints water, 1 dessertspoonful of sugar, 2 teaspoonfuls of salt, ½ teaspoonful carb. soda.

Electroculture Soup

A vegetable soup as satisfying and nourishing as that made with meat. Take 3 turnips, 3 parsnips, 3 carrots, 1 onion and ½ pound of split or shelled green peas, cut up vegetables small, put in saucepan with enough water to cover, salt and pepper to suit; boil for 2 hours; makes a pot of delicious soup for small family.

Lightning Source UK Ltd.
Milton Keynes UK
UKHW040041271122
412773UK00019B/331